FLY, FLY DRAGONFLY

How Scientific Data Tells a Story of Dragonfly Migration

Written by
Mary Bahr

Illustrated by
Jenny Dearinger

Photos by
Mary Bahr, Cathey Perkins, & Dan Jackson

Fly, Fly Dragonfly

How Scientific Data Tells a Story of Dragonfly Migration

Written by Mary Bahr
Illustrated by Jenny Dearinger
Photos by Mary Bahr, Cathey Perkins, & Dan Jackson

Everfield Press
Newberry, Florida

ISBN: 978-1-946785-91-6
LCCN: 2026936718

Fly, Fly Dragonfly

How Scientific Data Tells a Story of Dragonfly Migration

As the story begins, you will find opposite pages that tell nature's migration story through Lulu, the common green darner *(Anax junius)*, on the left, and, on the right, the scientists' story of how we know it.

Nature in all its complexity and beauty is probed, counted, and reduced to columns of numbers and facts by science. The result, when you interpret that data, is an endlessly unfolding story about things we never understood before, including how dragonflies migrate.

You will also learn about other migrating dragonflies and a young scientist's research on another migrating dragonfly, the Canada darner.

The last section provides advice to young citizen scientists on how they can contribute to the study of migration.

Common Green Darner Anatomy

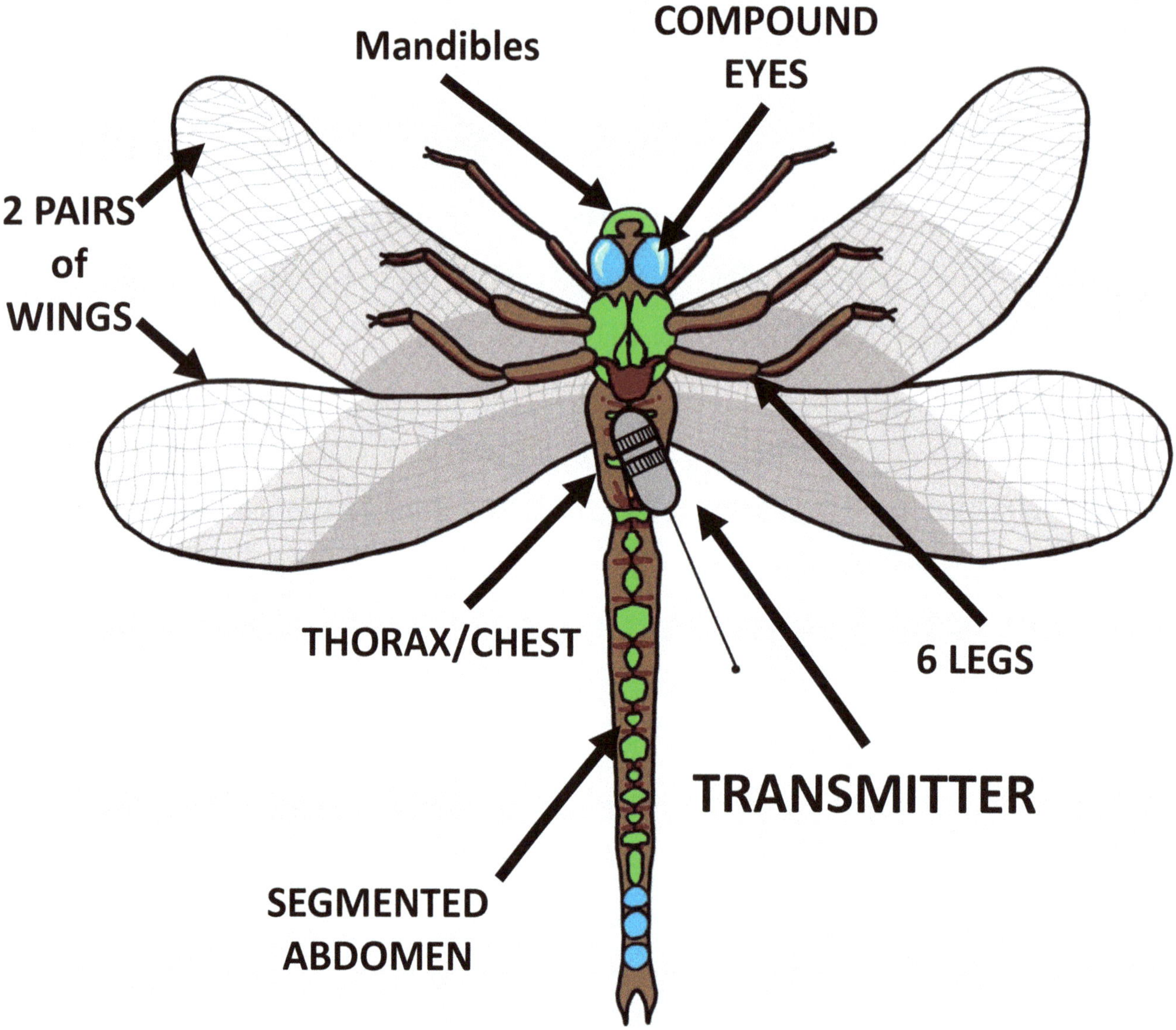

Dragonfly

We fly down the ridge
to the shore, following the
paths of our ancestors.

Our flying skills,
shaped by nature
over millions of years,
are written into our
muscle and brain.

Migrating
south for the winter.
We have been kings of the air
far longer than the humans
below us have lived on Earth.

Male common green darner *(Anax junius)* in flight.
Photo by Cathy Perkins.

Scientist

Standing on a New Jersey beach, I watch for dragonflies, common green darners: the green and blue males, and gold and green females all flying south in the autumn sun.

I am an **entomologist**, a scientist who studies insects. I call common green darners by their scientific name: *Anax junius*.

Dragonfly fossils show they have lived on Earth for more than 300 million years,
one hundred times longer than my **species**.

Dragonfly

Humans with nets
chase us.

My wrap-around eyes see
the net coming in slow
motion. I calculate its
path and fly away.

I twist and turn in the
sun desperate to escape.

At last, a human
captures me.

Scientist

Dragonflies are tough to catch. **Compound eyes** help them see in almost every direction. They see faster than humans, and their brains can predict where a mosquito or a net will be. We watch one fly backward out of my net. But we know they have a blind spot where their eyes meet at the back of their heads. I swing from behind. Success! I catch a gold and green female and name her Lulu after my great-grandmother who was a beauty too.

Photo by Mary Bahr.

Lulu the Dragonfly

The human pins my
wings together.
It does not hurt but
the urge to escape is
overwhelming.

I vibrate my muscles
trying to fly.

I am turned upside
down.
There is a strange
weight
stuck to my chest.

Scientist

Holding Lulu's wings gently but firmly, I turn her upside down. I use a strong glue to attach a miniature **radio transmitter** on the underside of her thorax.

Many scientists think that dragonflies migrate, but we need scientific evidence. We need **data**! The radio transmitter will record data on Lulu's location for the next ten days.

Photo with permission from Dr. Martin Wilkelski.

They let go of
my wings.

I perch
On a strange
warm twig.
I'm a little
dazed.

The weight on
my chest is
heavy.

But I am
powerful!

Scientist

We wait.
No one has ever done this before.
Can Lulu carry the weight of the transmitter?

The one-**gram** darner weighs a little less than a jellybean. A **milligram** is only a thousandth of a gram, but the 300-milligram transmitter is almost a third of Lulu's body weight.

She rests on my finger. Her wings vibrate.
She launches into the sky and flies like the transmitter isn't there!

Photo by
Mary Bahr.

Lulu the Dragonfly

Dragonfly,
Fly,
Fly!

Into
the
Sky!

I soar
Up,
Up!

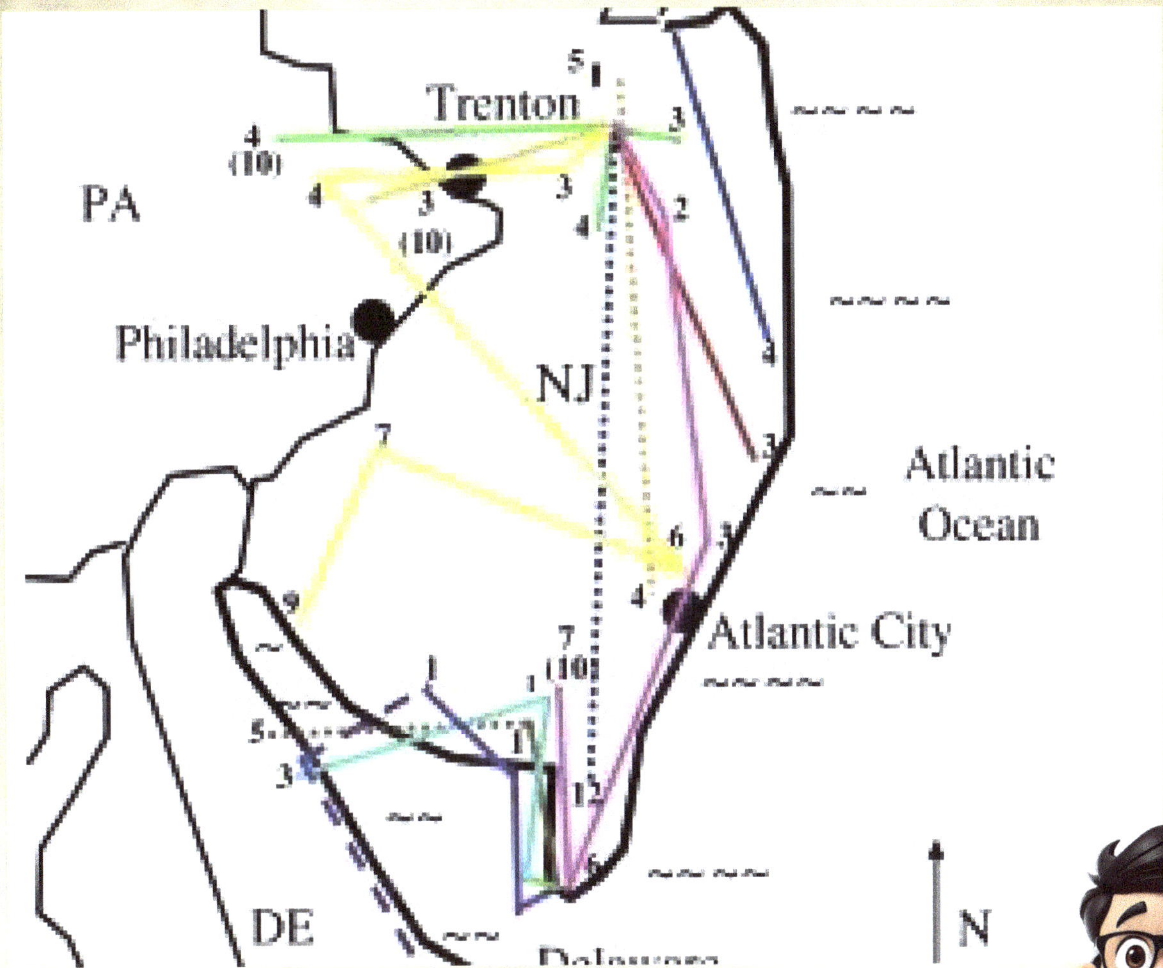

Each color tracks a different dragonfly.
Lulu followed the yellow line.

Scientist

Every day for ten days we track the transmitter data from Lulu and nine other common green darners we manage to catch:

- From the ground
- By car
- On foot
- In the air from a plane

Lulu the Dragonfly

We fly
out to sea,
back to the
shore,

moving steadily
south.

We cross the
Delaware River.

I lay eggs in a
little pond
along the way.

Migrating male common green darner *(Anax junius)* in north Florida.
Photo by Mary Bahr.

Scientist

We calculate flight distance using the data from the radio transmitters attached to our dragonflies.

They fly steadily south:

- One mile,
- Six miles,
- Fifteen miles,
- Sixty miles a day!

Lulu the Dragonfly

We follow the
weather,
hanging out when it is
warm,
doing our feeding
dance.

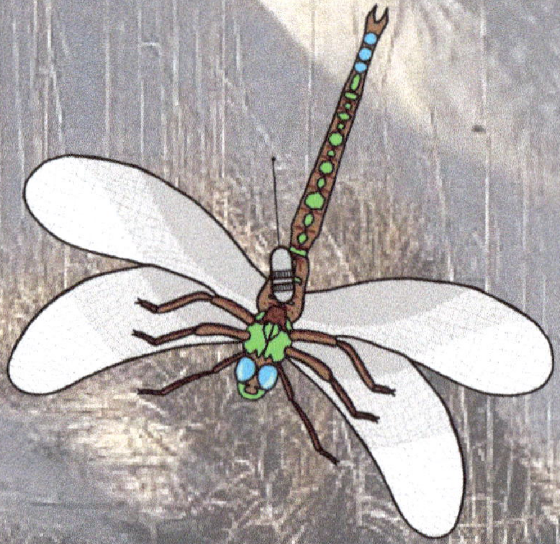

Eating flies and
mosquitoes
by the thousands.

We hurry south
when it turns cold.

Scientist

Our radio transmitter data tells us that migrating dragonflies fly on average every third day.

Male common green darner
(Anax junius) in north Florida.
Photo by Mary Bahr.

We combine our transmitter data with weather reports. The green darners always migrate after cold nights, resting and feeding after warm ones.

We track them for 10 days:

- Some fly far
- Some disappear
- All migrate

Our data predicts that in two months common green darners could fly 400 miles!

Lulu the Dragonfly

Hundreds of miles north
of the
New Jersey beach
where I was captured

my green darner relatives
fly by the millions
high above a Great Lake.

They are migrating south.

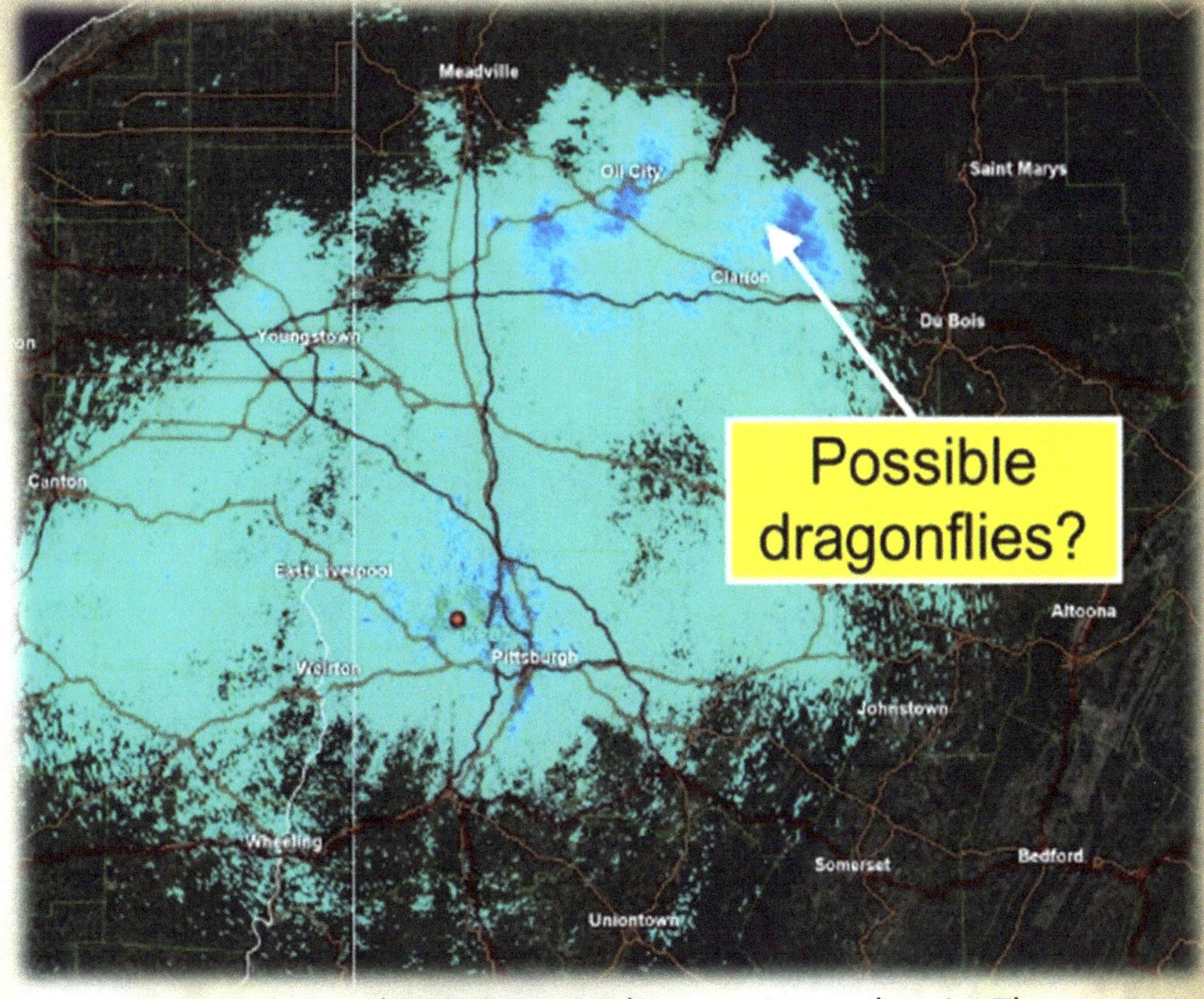

National Weather Service Radar over Pennsylvania. The jagged blue returns are probably migrating dragonflies.

Scientist

We see **radar** signals that suggest millions of dragonflies are migrating across Indiana, Ohio, and Pennsylvania along the south shores of the Great Lakes.

We know the radar signal is not rain because it is a clear day and because the signals have jagged edges instead of the smooth returns you see from raindrops. This new data helps support our **hypothesis** that dragonflies migrate.

Lulu the Dragonfly

As my relatives and I
fly across the country

We look down at
humans below us
looking up,

tracking us with
strange shiny tools,

watching us migrate.

Citizen Scientist

Friends at Paradise Beach on Lake Superior watch dragonflies fly by their sunset campfire. Their binoculars and cameras confirm they are green darners flying steadily south down the beach. A citizen scientist sitting by the fire knows this means these dragonflies are migrating.

Lake Superior north shore, Paradise Beach.
Photo by Mary Bahr.

She reports her sighting to the Migratory Dragonfly Partnership, which gathers observations from dragonfly watchers all over North America. Citizen scientists' sightings confirm the jagged-edged radar. The data really shows dragonflies migrating. Scientists use volunteer observations to verify indirect evidence like radar and satellite returns.

Lulu the Dragonfly

Young (called **nymphs**)
from eggs laid last year
by our grandparents

hatch from the waters
below us, and
join our migration.

The next generation
will winter in the deep
south, safe from the
cold.

Scientist

After the radio transmitter data confirmed that dragonflies migrate, we learned to trace adult dragonfly migration by measuring **hydrogen isotopes** from their wings.

Hydrogen Isotopes

A diagram of the 2 forms of hydrogen atoms. Count the number of particles in each nucleus.

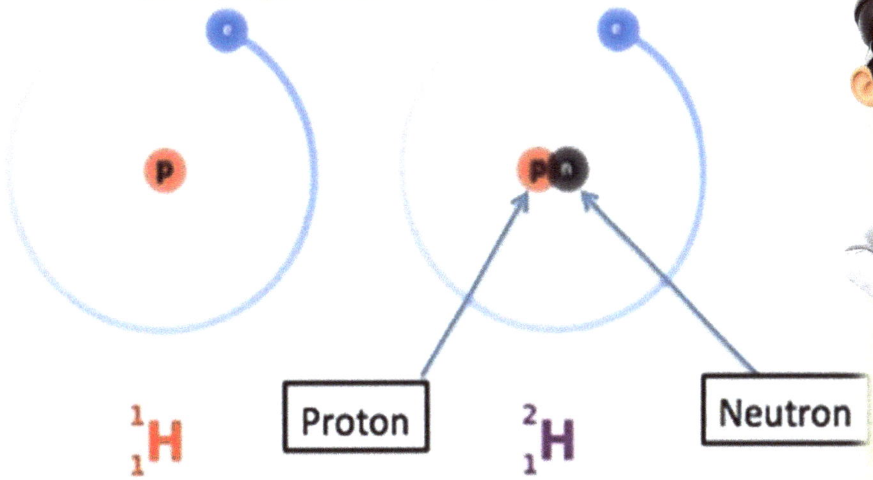

These slightly different forms of hydrogen **atoms** show us how far north each young dragonfly lived as an underwater nymph. The amount of heavier hydrogen atoms with two particles in the **nucleus** (called deuterium) decreases as you move north from the equator. The waters dragonfly nymphs call home add different hydrogen isotope combinations to their bodies as they grow.

We can map their movements by comparing where they were captured as adults with where the hydrogen fingerprints in their wings show they grew up.

Lulu the common green darner helped scientists understand the beginning of the dragonfly migration story.

Like monarch butterflies who take three generations to reach Mexico, the darners caught on a New Jersey beach are the first of a three-generation dragonfly migration.

They are also part of the first experiment that provided evidence for dragonfly migration.

Lulu's children will live south of the winter frosts and snow.
In the spring, her grandchildren will fly north again, leaving eggs along the way,
for the next year!

Scientist

*When we combine our **data** it tells a migration story*

Map of common green darner migration on the **US Eastern Seaboard**.

Like pieces of a jigsaw puzzle, hydrogen isotope data combines with citizen scientists' observations, radar returns, weather reports, and radio tracking to create a picture of a three-generation migration of common green darners. The first generation flies south. The second winters safe from the cold. The third flies back north in the spring. Or is it two generations? Some scientists think so. Part of the scientific process is looking for evidence to support competing explanations of the natural world. That is what is happening right now in dragonfly research. Science is never settled and scientific ideas can change when new evidence is uncovered.

Now that you have seen my migration story, you should know that many of my dragonfly cousins migrate too.

Here are a few of their migration stories, starting with a northern cousin, the Canada darner.

The human who studies them is a Minnesota scientist, Dr. Emily Schilling.

Another Migration Story

Photo reproduced with the permission of Dr. Schilling.

Hello! I am Dr. Emily Schilling, a freshwater ecologist. I taught at Augsburg College in Minneapolis, Minnesota, and now work for The University of Minnesota. I research dragonfly migration.

My earliest memories were playing in the stream that fed into Irondequoit Bay on Lake Ontario in upstate New York. I went from there to teaching sailing in the Caribbean. I have always been a “water person.”

I became interested in dragonflies started in graduate school when I studied the ecology of high-altitude lakes in Maine. These boggy lakes lack fish, giving dragonfly nymphs a refuge from one of their main predators.

These lakes are where I started thinking about dragonflies. I was very much thinking about them from the aquatic life stage perspective. You probably think of dragonflies as adults flying above us. They spend most of their life as nymphs in waters below us. For this reason, I first studied Canada darner migration in the upper Midwest by studying the underwater nymphs.

One spring, my students from Augsburg recorded when Canada darner nymphs hatch, and when we saw the adults flying. Observations tying events in nature to the seasons are called **phenology** and are useful when you study migration.

In central Minnesota Canada darner nymphs start hatching in late May and adults are flying by early June. In the summer, I and some of my students go north to catch adult Canada darners near the Canadian border.

We collect wing tissue for hydrogen isotope analysis to locate the adult's home ponds. The hydrogen isotope results, the location where the adult is captured, and the time of year combine data from chemistry, geography, and phenology data to give us a picture of the Canada darner's migration north in the spring.

I still want to explore if all Canada darners fly north or if some stay home in central Minnesota. Questions remain!

These dragonfly nymphs from a pond Emily studies are of different sizes. The largest ones will probably hatch this year while the smaller ones may overwinter in the pond and hatch next year. Dragonfly nymphs take anywhere from 30 days to 5 years to mature with the average being 1 to 2 years. Doctor Schilling calls Canada darners by their scientific name: *Aeshna canadensis.*

Photo reproduced with the permission of Dr. Schilling.

More Migrating Dragonflies

Carolina saddlebag
(Tramea carolina),
north Florida
Photo by Mary Bahr.

variegated meadow hawk
(Sympetrum corruptum),
Minnesota
Photo by Mary Bahr.

North American dragonflies that migrate include:

- multiple species of saddlebags (genus name *Tramea*)
- variegated meadowhawks (*Sympetrum corruptum*)
- twelve spotted skimmers (*Libellula pulchella*)

twelve-spotted skimmer,
(Libellula pulchella),
northern Minnesota
Photo by Mary Bahr.

These species have not been studied as intensively as the common green darners. But they are often observed migrating and are tracked by the Migratory Dragonfly Partnership.

A Migrating Champion

The star of migratory dragonflies is the wandering glider *(Pantala flavescens)*. Swarms of this small golden-brown dragonfly are common in the Southeastern United States in the fall.

Wandering glider captured from a migrating swarm in north Florida.

Photo by Mary Bahr.

This traveling dragonfly is found worldwide and holds the distance record for insect migration. Each year wandering gliders fly back and forth across the Indian Ocean, a distance of 14,000 **kilometers** or 8,700 miles. That's like flying from Sydney, Australia to Dallas, Texas!

Millions have been observed traveling toward Africa in October. Smaller numbers return to Southern India in May.

You Too Can Study Dragonfly Migration

If you enjoy nature, dragonflies are a fascinating and beautiful way to study it. And scientists like Emily can use your help observing dragonfly migrations.

A Young **Citizen Scientist** learns dragonfly identification. Photo by Mary Bahr.

The Migratory Dragonfly Partnership and books on dragonflies are a great place to start.

Getting Started, In-Hand Guides, and What Equipment You Will Need

When you see dragonflies migrating, they will be flying steadily south in the fall. You can enter observations online into the Migratory Dragonfly Partnership database using the protocols (scientific procedures) described on their website. You can also enter your observations into the iNaturalist database, which many scientists use in their research. The iNaturalist database is one of the largest available collections of data about the natural world, compiled by citizens and scientists.

Because there are over 300 dragonfly species in North America, look for a guide for your region, like Kurt Mead's *Dragonflies of the North Woods*, Gif Beaton's *Dragonflies & Damselflies of Georgia and the Southeast*, or a waterproof folding pocket guide, like James Kavanaugh's *Dragonflies & Damselflies: A Folding Pocket Guide to Familiar, Widespread North American Species*,.

Photos by Mary Bahr.

Some of Lulu's more distant relatives are the regal darner *(Coryphaeschna ingens),* above, much larger than Lulu, and an emerald spreadwing *(Lestes dryas)*, right, which is much smaller.

Equipment needed to start includes binoculars and a camera to record your observations. Green Darners fly fast and are very difficult to photograph on the wing. When they perch, they usually hang downwards from a tree trunk.

Male common green darner hanging from
a cypress tree in North Florida.
Photo by Mary Bahr.

As you begin to identify more dragonflies, a net and a hand lens are useful. Nets are available from biological supply companies. Any good ID guide will have introductory chapters on equipment and dragonfly biology and anatomy, which will be useful when you begin identifying all Lulu's cousins.

Common Green Darners are large and easy to recognize, even in flight. Kurt Mead describes their plain green thorax (chest area) and the large eyespot on the back of their head as distinguishing characteristics. Males will have a blue abdomen when mature, while the adult female has a green thorax and gray abdomen. Juveniles have a green thorax and a reddish-brown abdomen.

A mating pair of common green darners. The male with his blue abdomen is above and the female is below. The eyespot can be clearly seen on the back of each head.

This pair was flying through my garden sprinkler and was covered with water droplets.

Photo by Mary Bahr.

Another important part of dragonfly identification is behavior. For instance, a similar but smaller species widely found in green darner territory is the eastern pondhawk. The male looks similar to the green darner with a blue abdomen. But you will usually find these dragonflies near a pond's surface. They stay near their ponds and do not fly as fast and far as the similar-looking male green darner.

A male pondhawk *(Erythemis simplicicollis)* perched near the water surface in north Florida.

Photo by Mary Bahr.

When you observe dragonflies, questions may occur to you. And your research may find that there is no answer to your question yet. This is where the scientific process begins, just as it did several decades ago when Dr. Martin Wilkowski glued the first radio transmitters to his captured green darners to find out if dragonflies migrate. That was his question.

The author wonders if green darners hatch a migrating generation, as she has observed that beginning migrators on Lake Superior's north shore are all juveniles.

What are your questions about dragonfly migration?

Science Glossary for *Fly, Fly Dragonfly*

This glossary explains the highlighted science words in the text.

Atom - A tiny particle that makes up everything around us. Atoms contain smaller pieces in their centers, called a nucleus, containing protons (which have a positive charge) and neutrons (with no charge). Another tiny particle with a negative charge, called an electron, fills the space around the nucleus. The scientists use hydrogen atoms to help them study dragonflies.

Compound Eyes – Many tiny eyes combine to form larger ones that often wrap around an insect's head, giving them 360-degree vision.

Compound eyes on a Canada darner with 30,000 tiny eyes (facets).

Photo by Mary Bahr.

Data - Information, like facts, numbers, or descriptions, that we can collect and use to learn and to make decisions to test a hypothesis. Data in *Fly, Fly Dragonfly* includes radio transmitter data, radar returns, hydrogen isotope levels, and citizen scientist observations.

Gram - A tiny unit to measure how much something weighs. A gram weighs the same as a small paperclip, a few grains of sand, or a large dragonfly. Lulu weighs one gram.

Entomologist – A scientist who studies insects.

Hypothesis - An idea that can be tested based on the evidence collected from scientific experiments and observation. The hypothesis for the first experiment in *Fly, Fly Dragonfly* was that dragonflies migrate south in the fall.

Science Glossary for *Fly, Fly Dragonfly* (continued)

Isotope – Atoms that make up the same kind of substance may have slightly different weights because they have different numbers of neutron particles in their nucleus. For example, hydrogen always has 1 positively charged proton, which makes it hydrogen. But it can have either zero or one neutron to make two different types of hydrogen called isotopes. Deuterium, a hydrogen atom with two neutrons, is heavier than regular hydrogen, which has only one neutron.

Kilometer - A kilometer (km) is a unit used to measure long distances, like the distance between cities or towns. One kilometer is equal to a little over half a mile (or 0.621 miles, to be exact).

Migration – The act of traveling from one region to another. Migration usually happens with the seasons. In the northern hemisphere, most animals that migrate move south in the fall and north in the spring. The dragonflies in *Fly, Fly Dragonfly* do this.

Milligram –A tiny unit of weight, like a small grain of sand or a snowflake. A thousand milligrams make a gram. The radio transmitter in *Fly, Fly Dragonfly* was a third of a gram or 300 milligrams.

Nymphs – Young insects that shed their skins a number of times before they become adults. If the adults have wings, nymphs will have no wings or incomplete ones. Examples of young insects who are nymphs include grasshoppers, cockroaches, and dragonflies. Dragonfly nymphs live in the water and crawl out to transform into adults with wings.
(See the photo on page 26.)

Science Glossary for *Fly, Fly Dragonfly* (continued)

Odonata – A group of insects made up of dragonflies and damselflies. Dragonflies are larger than damselflies and are the only insects in the Odonata group that migrate.

dragonfly

damselfly

Phenology – The study of nature's calendar, recording data like the connection between when ice goes out on a lake and when dragonfly nymphs emerge from the water and become adults..

Radar - A radar system sends out radio waves like a spotlight. These waves travel until they hit something, like an airplane. The waves then bounce back, like an echo, to the radar system. By measuring how long it takes for the waves to return, the radar can tell you how far away the object is and how fast it's moving. Radar returns can show migrations of flying birds and insects.

Radio transmitter – A special machine that tracks its location by sending out invisible waves called radio waves. Radio transmitters can also send out sounds, music, or pictures (like TV) using invisible radio waves.

Species - A group of similar animals or plants that can have babies with each other, and those babies can also have babies. For example, common green darners mate and lay eggs that hatch into nymphs which produce more adult common green darners who then go on to make more common green darners. All the migrating species in *Fly, Fly Dragonfly* only produce their own kind. A Canada darner would never lay eggs that hatch to become a common green darner because they are different species.

Dr. Emily Schilling in the field catching dragonfly nymphs from their underwater home.

One of Lulu's cousins, a female lake darner *(Aeshna eremita)*, momentarily perched on a 'warm twig' attached to the author's hand.

The End...

...of the dragonfly migration story.
(For now!)

www.ingramcontent.com/pod-product-compliance
Lightning Source LLC
LaVergne TN
LVHW071346210926
848486LV00003B/239